Shapes

David Kirkby

RIGBY
INTERACTIVE
LIBRARY

© 1996 Rigby Education
Published by Rigby Interactive Library,
an imprint of Rigby Education,
division of Reed Elsevier, Inc.
500 Coventry Lane,
Crystal Lake, IL 60014

Cover designed by Herman Adler Design Group.
Designed by The Pinpoint Design Company
Printed in China

00 99 98 97 96
10 9 8 7 6 5 4 3 2 1

Library of Congress Cataloging-in-Publication Data
Kirkby, David, 1943–
     Shapes / David Kirkby.
       p.   cm. — (Mini math)
     Summary: Uses simple activities to explain shapes,
including circles, squares, rectangles, spheres,
pyramids, and cones.
     ISBN 1-57572-002-7 (lib. bdg.)
     1. Geometry—Juvenile literature.   [1. Shape.]
I. Title.   II. Series: Kirkby, David, 1943– Mini math.
QA445.5.K572   1996
516—dc20   95-38717
                                     CIP
                                     AC

Acknowledgments
The publishers would like to thank the following for the
kind loan of equipment and materials used in this book:
Boswells, Oxford; The Early Learning Centre; Lewis',
Oxford; W.H. Smith; N.E.S. Arnold; Tumi. Special thanks
to the children of St Francis C.E. First School.

The publishers would like to thank the following for
permission to reproduce photographs: J. Allan Cash Ltd,
p. 10; Robert Harding Picture Library, p. 20.
All other photographs: Chris Honeywell, Oxford.

contents • contents

A square has 4 straight sides, all the same length. It has 4 matching corners.

Squares tessellate.
They fit together with no gaps.

How many squares can you see?

• To Do •

Draw a house with
4 square windows.
Draw a big square.
Make 4 squares
tessellate inside it.

5

A rectangle has
4 straight sides.
It often has
2 long sides and
2 short sides.
It has 4 matching
corners.

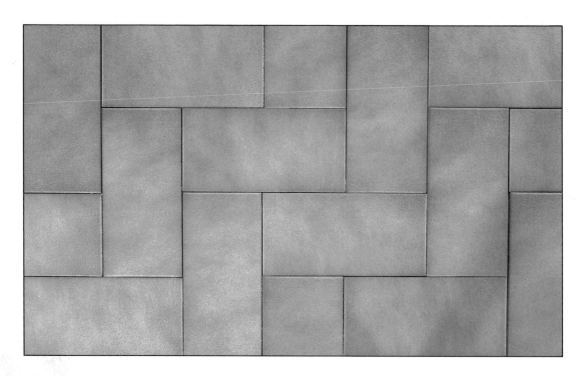

Rectangles tessellate.
They fit together with no gaps.

6

How many rectangles can you see?

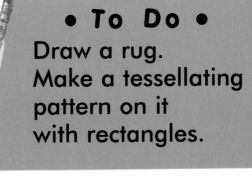

• To Do •

Draw a rug.
Make a tessellating
pattern on it
with rectangles.

A triangle has 3
straight sides.
The corners do
not need to match.

This triangle has sides of the
same length.
Some triangles have sides of
different lengths.

How many triangles can you see?

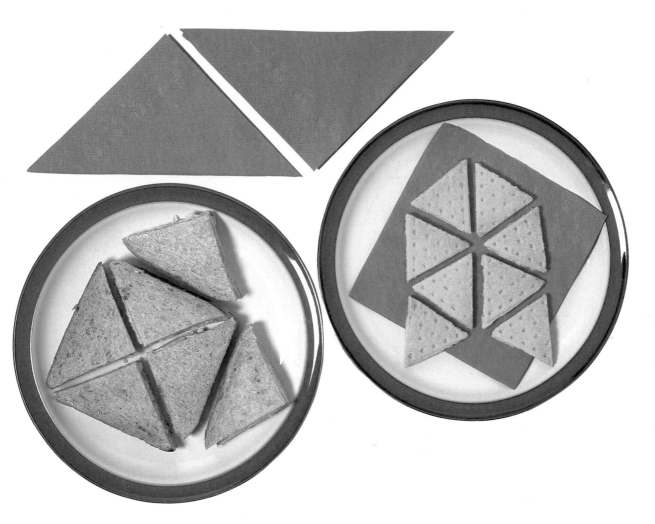

• To Do •

Draw the fattest
triangle you can.
Draw the thinnest
triangle you can.

These are circles.
A circle has no
corners.

This target has
circles inside circles.

How many circles can you see?

• To Do •
Draw some circles.
Do circles tessellate?

A cube is a solid shape.
It is not flat.
It has 6 faces
(sides).
Each face is a
square.

Some food is cube shaped.

How many pictures can the cube
puzzle make?

• To Do •

Copy this shape.
Cut it out and make
a cube.

A rectangular prism
is a solid shape.
It has 6 faces (sides).
These faces can be
rectangles or squares.

These rectangular prisms
are all the same size.

How many rectangular prisms can you see?

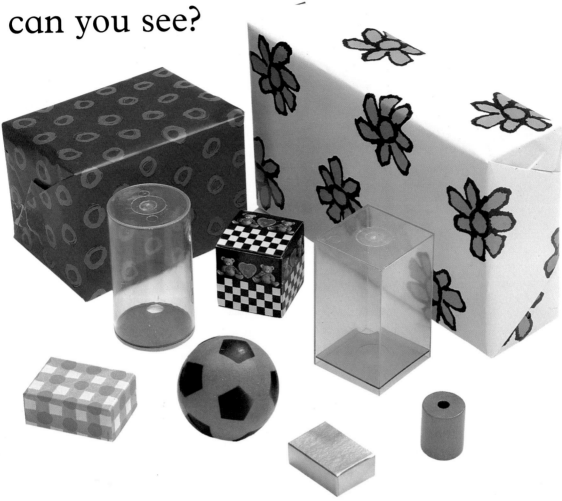

• To Do •

Find a rectangular prism. How many faces does it have? Are the faces all the same size?

A cylinder is a solid shape.
It is tube shaped.
The end faces are circles.

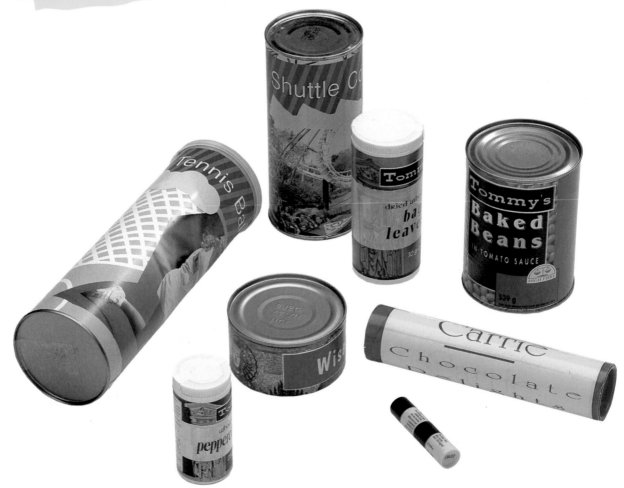

These are all cylinders.

How many cylinders can you see?

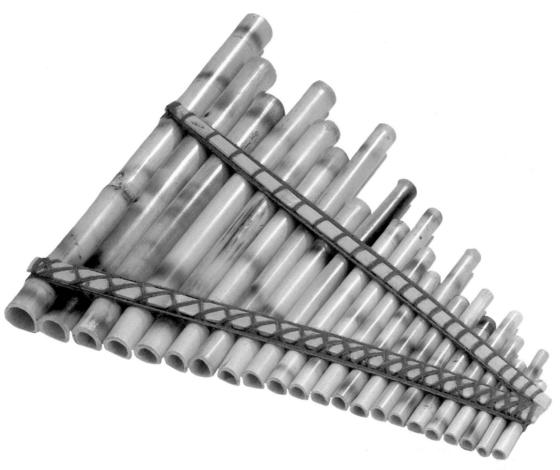

• To Do •

Can you make a tower with cylinders on their ends?

What happens when you try to make a tower using cylinders on their sides?

A sphere is a solid shape.
It is round.
A ball is a sphere.

A sphere is good to roll.

How many spheres can you see?

• To Do •

Can you make a tower from cubes?
Can you make a tower from rectangular prisms?
Can you make a tower from spheres?

A pyramid is a solid shape.

It has triangle-shaped sides that join at the top.

It can have a different shape for a base (bottom).

Pyramids can be very big.

How many of these pyramids
have square-shaped bases?

• **To Do** •

How many sides
does a pyramid with a
square base have?
How many sides
does a pyramid with a
triangle base have?

A cone is a solid shape.
It has a circle for a base and comes to a point at the top.

These are all cones.

The tops of these are cone-shaped.
How many cones can you see?

• To Do •
Copy this shape onto
paper or card.
Cut it out.
Make a cone shape.

GLUE

answers • answers

Page 5   10 squares

Page 7   5 rectangles

Page 9   16 triangles

Page 11   3 circles
          To Do: No, circles do not tessellate.

Page 13   6 pictures

Page 15   6 rectangular prisms

Page 17   22 cylinders

Page 19   7 spheres
          (The rest of the bubbles are not perfectly round.)

Page 21   1 pyramid
          To Do: 4 sides, 3 sides

Page 23   9 cones